SURVIVE ANY EMERGENCY

From Blizzards to Terrorism

Betty-Anne Lawlor

Chestnut Publishing Group

Library and Archives Canada Cataloguing in Publication

Lawlor, Betty-Anne, 1959–
    Survive any emergency : from blizzards to terrorism / Betty-Anne Lawlor.

ISBN 1-894601-36-X

    1. Survival skills. 2. Emergency management. I. Title.

HV551.2.L39 2005   613.6'9   C2005-906125-1

Cover and design by John Zehethofer
Typesetting by Laura Brady

Printed and bound in Canada.

Published by Chestnut Publishing Group
4005 Bayview Ave., Ste. 610
Toronto, ON M2M 3Z9 Canada
Tel: 416 224-5824 Fax: 416 486-4752
www.chestnutpublishing.com

We acknowledge the financial support of the Government of Canada through the Book Publishing Industry Development Program (BPIDP) for our publishing activities.

Dedicated to:

My idol:
Mom,
you always loved, believed, supported,
and guided me.
I honor your moral integrity and
I strive to be just like you.

Dad, thank you for a strong and determined character.
I love and miss you always.

My children:
Bobby, James, Andrea, and Caterina,
you are my life's reward.

Auntie Elaine and Auntie Marie:
Thank you for all the love and
support you have given me over the years.
They will always be treasured memories.

Special thanks and acknowledgement for time given, editing, assistance, and encouragement go to:

Mr. Jean-Bernard Guindon, Director of the Emergency Preparedness Center, City of Montreal; Mr. James P. Lawlor, President Lawlor Holdings Inc.; Dr. Savas Fortis; Dr. Wendy Krasny; Maitre Martin D. Charest, LLB; Ms. Susan Ehret, Office of the Vice President, External Relations, Concordia University; Mrs. Yolanda Nower RNA, Montreal General Hospital; Mrs. Diana Peck; Mrs. Colleen Hoffman, Author; Ms. Julie Vien; Mr. Robert Lawlor-Steger; Ms. Eva Perlman; Mrs. Nina Avraam-Rodoussakis; Mr. Stanley Starkman and Mr. Harry Goldhar for taking a chance and believing in me.

Contents

Introduction *1*

PREPARATIONS

1. Home–staying or leaving *5*
2. Care for the elderly and physically challenged *9*
3. Tips for pet owners *11*
4. Important telephone numbers *13*
5. Emergency kits *15*
   – food *15*
   – water *17*
   – first aid *20*
   – medicine *20*
   – vehicle *22*
6. Personal packing list *25*
7. Important miscellaneous items *27*
8. Finances *29*

EMERGENCIES

9. Earthquakes *33*
10. Tsunamis *37*
11. Floods *41*
12. Hurricanes *43*
13. Tornadoes *45*
14. Hail, thunder, and lightning storms *47*
15. Landslides *51*
16. Heat waves *55*
17. Blizzards and ice storms *57*
18. Fire *63*
19. Chemical and biological emergencies *69*
20. Terrorism *75*
21. If disaster strikes *81*
22. First aid *83*
23. General safety procedures following disasters *89*
24. Emotional reactions *93*
25. Bi-annual checklists *95*

Introduction

I have searched far and wide looking for information about emergency and evacuation procedure guidelines. A few sources have helped, but none were in-depth enough to satisfy me. I believe we are at a point when we must count on ourselves to protect and look after our families and friends. As a result, we must plan for the possibility of some serious situations heading our way.

These situations need not cause panic, and we *can* get through them. We cannot control the actual event, but we can prepare for many different types of emergencies. Your and your family's welfare and safety are worth the effort!

Some potential emergency situations are weather-related, such as winter storms, floods, fire, earthquakes, tornadoes, etc. Others are man-made, such as chemical emergencies, epidemics, and terrorism.

Emergency services in your area will not be able to help everyone. Their priorities will be the handicapped and the wounded. The general population will be considered as a group, and will be treated as such. Emergency services cannot cater to our special or specific individual needs: it is up to us to take care of those needs in a crisis.

Remember, disasters can happen any time and anywhere, and you may not have much time to respond. Be prepared.

Betty-Anne Lawlor

Preparations

The information here is for optimum preparedness.
Not everyone will do it all,
but the more you do the better prepared you will be.

1 | Home — Staying or Leaving

Involve all family members in planning for an emergency. They must fully understand what needs to be done and what their role is in an emergency. The first way to be prepared is to take a first aid training course. It will teach you how to minimize the effects of injury and will help you to stay calm in a crisis.

Then start with yourself: have a tetanus shot or a booster if you haven't had one in the last 10 years. In addition, be alert and informed. Find out which local radio and TV stations broadcast emergency instructions and monitor them regularly. Know the warning systems your local government uses.

Be familiar with your workplace emergency plan and your children's school or day care emergency plans. Have written back-up care arrangements in place at the school or day care in the event that you are detained in an emergency. Employers, like schools, need up to date information about your medical needs and how to contact your family.

Ensure that elderly family members who do not live with you are included in your emergency plan. If they have special needs, register them in advance for special transportation or other special assistance programs, home care programs, or local seniors' organizations. See the chapter on care for the elderly, page 9.

Post all emergency plans and phone numbers in prominent places at home and in the office. You may wish to use the sheet on page 13 to keep track of the numbers. Select pre-determined meeting places if fire or another event forces your family out of your home.

Ask someone outside your immediate area to act as a central point of contact for your relatives and friends in an emergency. It will often be easier to call outside the province/state than within the affected area. In addition, e-mail, which uses different microfiber lines, is often more accessible than telephones (as long as the telephone poles and wires are up and running). The phone lines can get overloaded quickly. Cellular phones are wireless. As long as the towers are standing, they will work, although it is possible that the lines could be overloaded at certain moments.

In an emergency you may have to leave your home before all your family members arrive, and you may be unable to meet at the pre-planned designated place. Decide in advance where to leave a note – for example, in the mailbox or on the kitchen table – to explain where family members can meet you. Include information in the note about whether they need to take their backpack with them, or if you have them.

Ensure that all members of the family carry a form of identification at all times: for example, clothing label, wallet card, bracelet with name and address. A family photo is also a good idea. Pictures of your belongings and furnishings are useful for insurance claims.

Have your vehicle serviced regularly. Keep it filled with gas; do not let the gas level fall lower than half a tank. If the electricity supply is interrupted, the pumps at the gas station will not work. Driving routes should include an alternative route in case main roads are cut off or backed up and impassable. However, you should follow all directions that are specified by emergency personnel first.

Staying in your home during a disaster:
Depending on the nature of a crisis, you may have the option of stay-ing put or leaving, or you may have to be evacuated. If you plan to stay at home, or in case you are forced to stay there, conduct a hazard check of your home. Review safety features such as smoke detectors and how to monitor them. Look for frayed electrical cords or over-loaded circuits. Don't keep flammable materials near electrical equip-

ment or your furnace. Learn how to safely turn off the main water and electrical switches in your home.

Depending on the situation, you may lose power. You can purchase electric security lights that light up automatically if the power goes out. Be sure to have all the kit lists completely prepared (see pages 15–23 for various kit lists). Listen constantly to a battery or crank-operated radio (in case you suddenly lose power) for updates and other specific preparation tips. Follow the instructions for specific emergencies provided in later chapters.

Leaving your home during a disaster:
If you leave your home, do not assume an evacuation will last only a few hours. Plan to evacuate with enough items to keep your family comfortable for at least three to five days. Take one pre-packed duffle bag or backpack per person, one bag for first aid and medicine kits, and a bag for miscellaneous items. Children's activity items may be carried in their school bags. Double the quantities for children under 2.

If an emergency is imminent or if a disaster has already occurred, keep the phone lines open for use by emergency workers. Monitor local radio broadcasts for instructions and current information. Officials will advise if it's necessary to turn off main water and electrical switches. If so, and you cannot get access to the electrical switches, pull out all wall plugs for TV, stereo, radio, lamps, computer, kettle, toaster, microwave oven, fridge, stove, washing machine, dishwasher, and any other electrical appliance you have. Make a list so you don't forget any. If you have food in the fridge or freezer, and are able to, leave the power on. Close all windows and make sure to lock all doors, including garage and shed.

Your local area may set up an emergency reception center to provide food and shelter. Even so, bring your own short-term quantities of food with you, as the centers will be stretched thin until provisions and supplies arrive. Bring a kettle, an electric single cooking ring, and an extension cord. If the shelter approves and has electricity available, you will be able to make tea, coffee, soup, etc.

If you are leaving home, advise the emergency reception center, local government, or police of your destination.

REVIEW

- Take a first aid course.
- Have your tetanus/booster shots updated.
- Know your local warning systems, plus local TV and radio broadcast information stations.
- Familiarize yourself with the emergency plans for your office, school, and day care.
- Prepare your home emergency plans including: phone numbers, identification, contacts, driving routes, and evacuation. Post them in several places, including at work.
- Prepare your home for staying during the emergency:
  - heating equipment
  - cooking equipment
  - food supplies
  - water
  - battery or crank operated radio.
- Prepare your family for evacuation:
  - backpack
  - food and water
  - medical and first aid kits
  - last-minute check list for turning off appliances, electricity, water, and closing the home.

2 | **Care for the Elderly and Physically Challenged**

Many elderly and physically-challenged people live alone. Special provisions must be planned and prepared for them. Register them in advance for special transportation and with special assistance programs, home care programs, and local seniors' organizations. In addition, prepare their kits as for everyone else. Keep some emergency supplies beside the bed.

Anchor special equipment such as telephones and life support systems. Fasten tanks of gas, such as oxygen, to the wall. If electricity is required to operate the equipment, buy a generator to use in case of a power outage. If the person has a hearing difficulty, install a fire alarm system with flashing strobe lights that will attract his or her attention. Make sure there are extra batteries for hearing aids. Have the person keep walking aids near him/her at all times. Have extra walking aids in different rooms of the house. Put security lights in an outlet in every room in the house. Make sure the elderly or infirm have whistles in several prominent places in the house in case they have to signal for help.

If the person is a non-English speaker, prepare an emergency card, written in English and his/her native tongue, that lists the person's name, address and any special needs. Include emergency contact information for your family. Find two people who can be relied upon to check on the elderly or infirm during and after a crisis. Show them where the emergency supplies are and explain the special needs, or show them how to operate any equipment needed. Give them a spare key.

If an earthquake occurs be especially careful of those in bed or sitting down. They should stay put and cover their head and neck, provided nothing is likely to fall on them; the shaking could knock them down if they try to stand. If they are in a wheelchair, they should go to a doorway with no door that might swing and hit them.

REVIEW

- Plan and prepare special provisions.
- Register for special assistance.
- Prepare kits and emergency instruction cards.
- Anchor and secure special equipment.
- Install specialized alarms and security lights.
- Arrange for back-up people who can assist during an emergency.

3 | Tips for Pet Owners

Depending on the type of disaster, different responses must be planned. If you must evacuate, you must also evacuate your pets. Leaving pets behind, even if you try to create a safe place for them, is likely to result in injury, in animals getting lost, or worse. Do not count on local animal shelters, as they may be overburdened by animals that they already have and those displaced and abandoned during a disaster.

If you must leave your pets wherever you have found shelter, do your best to visit them daily. They will be scared and insecure in a strange place.

It may be difficult to find a shelter for your animal during a crisis, so plan ahead. Contact hotels and motels outside your immediate area to check if they will accept your pet. Keep a list of pet-friendly places, including phone numbers and directions, with other disaster information and supplies. Ask friends and relatives (if they are outside your immediate area) if they would be able to mind your pet during an emergency. If you have more than one pet, they might do better by being kept together, but do not count on that. Some people may not be able to care for more than one animal.

Prepare a list of your pets' necessities. Make sure their collar ID tags and vaccinations are up-to-date. Have at least a week's supply of food, with instructions stating quantity and times for feedings. Have water, bowls, a can opener, cat litter and pan, sturdy leashes and collars, bed, toys, and a pet first aid kit in a spare duffle bag ready to grab.

Make an instruction list. Put it in a plastic cover and tape it to the cage or carrier, or place it in the duffle bag tagged with the pet's name, address, and phone number. Include the veterinarian's phone number and address. List medications, vitamins, and medical records. Have a picture of the pets included in the pack in case they get lost. Advise if the animals have behavior problems.

Depending on the type of crisis, do not try to hold onto your pet. Animals will instinctively protect themselves and hide where they feel safe. If you get in their way, even the most placid of pets may turn on you. If you have outdoor pets, keep them inside until after the crisis has passed. Be patient with your pets after a crisis. Animals become stressed, just like people, and need time to settle down. They may disappear for a time, but will return when things are calm.

R E V I E W

- Update vaccinations and collar ID tags.
- Prepare food, supplies, and care instructions.
- List phone numbers, addresses and directions to pet-friendly places. Include this list in your general list.

4 | Important Telephone Numbers

Out-of-province/state contact:
Name: _____ City: _____

Telephone day: _____ evening: _____ cellular: _____

Local contact:
Name: _____ Telephone day: _____

evening: _____ cellular: _____

Nearest relative:
Name: _____ City: _____

Telephone day: _____ evening: _____ cellular: _____

Family work numbers:
Mother _____   Father _____

EMERGENCY TELEPHONE NUMBERS:

In a life-threatening emergency dial 911 or the local emergency services system: _____

Police dept.: _____

Fire dept.: _____

Ambulance: _____

Hospital for adult emergency room: _____

Hospital for children's emergency room: _____

Poison control center: _____

Adults' doctors: _____ _____

Children's doctors: _____ _____

Neighborhood pharmacy: _____

24-hour pharmacy: _____

5 | Emergency Kits

FOOD KIT LIST:

Store at least a three- to five-day supply of non-perishable food. If there is a possibility of a military attack or some other large disaster, you will need to prepare long-term food supplies. Select foods that require no refrigeration. Choose ones that require little preparation or cooking, and that require little or no water. In addition, try to select foods that are compact and lightweight. Freeze-dried and air-dried foods can be purchased from a sporting or camping equipment store. This will be your best form of stored meat. Don't forget to prepare foods for infants, elderly persons, or for those with special diets. Rotate your stored food every six months, or use by the expiry date and replace with fresh.

Include a selection of the following foods in your food kit:
- Meal replacement drinks (for example, Ensure, Boost)
- Frozen meat loaf (previously prepared and stored in the freezer)
- Ready-to-eat canned meats, stew, spaghetti, soups, and puddings
- Instant soup (if powdered, include extra water)
- Dried fruits and vegetables
- Juice (boxes/bottles/cans)
- Powdered and/or long-life milk, honey, frozen muffins (for example, banana & blueberry, banana & bran)
- Staples: almonds, tea bags, coffee, sugar, salt, pepper, seasoned salt

- High energy: peanut butter, jam/jelly, crackers, granola bars, trail mix, dehydrated fruits
- Comfort foods: Gum, hard candies, mints, chocolate bars, cookies, sweetened cereals.

Storage tips: Keep food in the driest and coolest spot in the house, a dark area if possible. Keep food covered at all times. Open food boxes or cans carefully so that you can close them tightly after each use. Wrap cookies and crackers in plastic bags and keep them in tight containers. Empty the contents of opened packages (for example, sugar, dried fruits and nuts) into screw top or airtight containers to protect them from pests. Inspect all food containers for signs of spoilage before use. Keep a few gel ice packs frozen; they are good for power outages and food transport.

Power outage: If the electricity goes out, use perishable foods and foods from the refrigerator first, then use foods from the freezer. To minimize the number of times you open the fridge or freezer door, post a list of fridge and freezer contents. That way you can keep the food cold longer instead of opening the door and looking each time to see what's in there. In a filled, well-insulated freezer, foods will usually still have ice crystals in their centers (meaning food is safe to eat) for at least three days. Begin to use non-perishable foods and staples last.

How to cook if the power goes off: For emergency cooking you can use a fireplace. A charcoal grill or camp stove is to be used outdoors only. You can also heat food with candle warmers, sterno, chafing dishes, and fondue pots. Local sporting stores carry several types of food warming methods. Canned food may be warmed and eaten right out of the can. Be sure to open the can, so the can doesn't explode, and remove the label first so the label doesn't catch fire.

REVIEW
- Prepare a three- to five-day food supply.
- Prepare foods for infants, elderly persons, and those with special diets.

- Prepare storage containers and have a specific storage area in the home.
- Prepare lists of freezer and refrigerated foods.
- Keep the list up-to-date.
- Prepare cooking/warming equipment.

Water

Stocking water reserves and learning how to purify contaminated water should be among your top priorities. A normally active person needs to drink at least 2 liters (4 pints) of water each day. Hot environments and intense physical activity can double that amount. Children, nursing mothers, and ill people will need more. A three- to five-day supply of 4 liters (8 pints) per person/day is required, half for drinking and half for food preparation, hygiene, and sanitation.

Store water in plastic containers such as soft drink bottles. Do not store in containers that will decompose or break, such as milk cartons or glass bottles. Never use a container that was previously used for toxic substances; tiny amounts of toxins may remain in the container's pores even after washing. You can also purchase food-grade plastic buckets or drums. Before storing your water, treat it with purification tablets or a preservative such as chlorine bleach to prevent the growth of micro organisms.

There are several ways to purify water. None are perfect. For example, you can let the water sit so that floating particles settle to the bottom, pour off the top, then strain the water from the bottom through layers of paper towels or clean cloths, and then purify. Boiling water is the best way to purify; boil furiously for about 10 minutes, allow it to cool, then pour it back and forth between two containers to put the oxygen back into it to improve the taste.

Distilling the water will remove microbes, heavy metals, salts, most other chemicals, radioactive dust, and dirt. To distil, fill a pot halfway with water. Tie a cup to the pot lid or pot handles and hang the lid upside down. The cup should not touch the water in the pot. Boil the water for 20 minutes. The water that drips into the cup from the lid is distilled water.

You may wish to use a couple of methods to clean water as much as possible. You can also use liquid bleach that contains 5.25 per cent sodium hypochlorite and no soap. Some containers of bleach warn "Not for personal use." Check before purchasing and treating the water.

If the water is not purchased and sealed with a long expiry date, rotate your stored, purified water supply every three to six months (depending on the purification method) so you always have a fresh supply.

If your supplies begin to run low during an emergency, never ration water. Drink the amount you need today, and try to find more for tomorrow. You can minimize the amount of water your body needs by reducing activity and staying cool. There are also hidden water sources in your home. Use water from your hot water tank, in your plumbing (familiarize yourself with its operation), and in ice cubes. As a last resort, use water in the reservoir tank of your toilet (NOT the bowl), but purify it first. In addition, waterbeds hold up to 400 gallons of water, but some waterbeds contain toxic chemicals that cannot be fully removed by many purifiers. If you designate a waterbed in your home as an emergency resource, drain it yearly and refill it with fresh water containing approximately 2 ounces of bleach per 120 gallons (verify this with the manufacturer).

To use the water in your pipes, let air into the plumbing by turning on the highest faucet in your house and draining the water from the lowest one. To use the water in your hot water tank, be sure the electricity, gas, and pilot lights are off, and open the drain at the bottom of the tank. Start the water flowing by turning off the water intake valve and turning on a hot water faucet. Do not turn on the gas or electricity when the tank is empty. Call your gas company to relight the pilot light on a gas-fired water tank.

If you hear reports of broken water or sewage lines during an emergency, shut off the valve for your incoming water supply to stop contaminated water from entering your home. If you do that, remember to shut off all water-using appliances and equipment.

Make sure water-using equipment is not exposed to temperatures below freezing. Water, when frozen, expands and can cause breaks that could damage the equipment. To avoid this damage, detach the unit at the bypass valve. Remove as much water as possible from the tank, then fill it with non-toxic antifreeze which can be purchased from any recreational vehicle dealership or hardware store. It is relatively easy to shut down most makes of equipment, but when you are not sure, call a professional. NOTE: If you rent your home, contact your landlord for permission or advice.

If you need to use water from outside your home, use the following sources: rain water, streams, rivers and other moving bodies of water, ponds and lakes, natural springs. Purify the water before drinking.

Avoid water with floating material, an odor or dark color. Use saltwater only if you distil it first. In addition to having a bad odor and taste, contaminated water may contain micro organisms that cause diseases such as dysentery, cholera, typhoid, and hepatitis. You should purify all water before using it for drinking, food preparation, or even personal hygiene if you are not sure of its purity.

REVIEW

- Learn how to purify water or stock prepared water reserves. Rotate the reserve every three to six months, depending on the purification method.
- Familiarize yourself with the hot water tank, plumbing systems, and other possible water sources in the home.
- Locate your home's main outdoor intake water valve and all the sinks' and water-using appliances' shutoff valves. Know how to turn them off.
- Purchase non-toxic antifreeze in case you have to shut off appliances or equipment that may freeze while shut down.
- Investigate your area for possible outdoor, drinkable water sources.

FIRST AID KIT LIST:

Try to be prepared for any kind of emergency. Many of the necessary items come in smaller sizes, perfect for travel. A smaller, separate kit should be made for teenagers, since they can become separated from you during a crisis. Put them in their school backpacks. Rotate use of the contents so they do not become outdated.

The following items can be put into a tight-seal plastic storage container or heavy duty freezer bags.

– First aid manual
– Regular, fabric, and antibiotic adhesive bandages
– Blister pads, ace bandage, and rubber bands
– Triangle slings, cotton swabs, cotton balls, pins
– Matches and lighter
– Burn/non-stick dressings, gauze pads, gauze wrap
– Tape, scissors, tweezers, needle and thread
– Nail clippers, emery board,
– Pen and paper
– Hand wipes, aloe gel, antibiotic ointment
– Calamine lotion, tincture of iodine
– Hydrogen peroxide, after-bite insect lotion
– Disinfectant hand gel, sore muscle rub, Vaseline
– Sunscreen SPF15 (minimum), sunburn ointment, latex gloves
– Dust masks, whistles, and swimming goggles.
– Mirror, for hard to see areas.

R E V I E W

• Prepare first aid kits for everyone, including a teenager's version, plus instructions.
• Rotate and use ointments and lotions once per year before the expiry date.

MEDICINE KIT LIST:

This is an important item. During an emergency you may not be able to reach an open pharmacy, it may not carry your particular brand of medicine, or you may not have the cash to spare during a crisis. Having an extra set of medicine need not be a waste, since you can rotate the medication every year with fresh supplies, and use the ones with the closer expiry date. Pack for a minimum of one month. Everyone should carry one set of their own medication or prescription, even to school (if they are responsible enough). You should also have individual multi-vitamins for the whole family.

Teenagers should have an age appropriate, separate emergency medicine kit since they can become separated from you during a crisis. Put the items in a freezer lock bag and store it in their backpacks. Write a contents list, including information about what each medication is for, the dosage, and the expiry date. For example:

Tylenol: for fever and pain, take 1 tablet with water, every 4 hours, do not exceed 8 tablets per day, expires March 2008.

Remember that you don't know what you may end up eating or not eating during an emergency. You may require assistance digesting or passing food.

To be prepared for anything, you need to pick brands ccording to your personal medical needs:
- Antihistamine, anti-diarrhoea medicine, menstrual cramp tablets
- Acetaminophen tablets and liquid, baby Tempra, antacid
- Anti-nausea tablets/liquid, cough syrup, allergy relief tablets/liquid
- Lip balm, eyecup and flushing liquid, syringe
- Liquid tear solution, mineral oil, stool softener/laxative
- Adult thermometer, child/baby thermometer, Isopropyl alcohol
- Children's suppositories
- Antibiotic eye/ear drops
- Mentholated chest rub, spoon, measuring cup.

REVIEW
- Prepare a one-month supply of medications and prescriptions, and include vitamins.
- Prepare a teenager's version of a medicine kit, including instructions.
- Rotate medicines once per year for use before the expiry date.

VEHICLE KIT LIST:

The following list may seem like a lot of stuff to carry around. It is not. It should all fit into a medium-sized plastic box (to keep everything dry). The plastic box can also double as a water catcher for hygiene, laundry, and drinking purposes (see water purification section, page 17). Store the box securely in the trunk of your car. Some stores have emergency-sized kits containing several of these items.

*Must-have items. The others are useful to have.

* Booster cables, windshield fluid and blades
* Portable storable gasoline
* Puncture seal gel in case of flat tire
* ABC fire extinguisher
* Emergency flares (to be placed a minimum 50 feet (15 meters) away from your vehicle to give advance warning to oncoming vehicles)
* Maps of your city, province/state, and surrounding provinces/states
– Phone adaptor kit for charging cell phones.

Winter items:
  – Methyl hydrate to de-ice the fuel line, motor oil, transmission oil
  – Power steering fuel, brake fluid, WD-40
  – Kitty litter or salt, lock de-icer, ice scraper, shovel.

The following items should be kept in your trunk in a box at all times:
  * First Aid manual
  * Basic first aid kit including:
    – 6 triangular bandages
    – 4 pressure dressings
    – 12 pieces 4 x 4 sterile gauze dressings
    – 1 large non-stick burn dressing
    – 1 3-inch Kling roller bandage
    – 1 roll of 1-inch adhesive tape
    – 36 adhesive strips
    – tissues

- pair heavy-duty first aid scissors
- 6 large safety pins
- Antiseptic towelets
- Needle, thread, buttons, shoelaces
- Latex gloves
- Notepad and pencil.
* Emergency blanket (weatherproof foil or plastic)
- Scotch tape and rubber bands
- Change for telephone calls from a phone booth
- Candles, waterproof matches, lighter
- Tin plates/cans to stand candles in
- Rope, bungee cords, heavy tape
- Garbage bags and ties
- Plastic pack of hand warmers
- Extra sunglasses, flashlight
- Nails, multi-head screwdriver
- Granola bars, crackers, cough drops
- Candies, mints, gum.

6 | Personal Packing List

Packing should be as compact as possible. Pack all items in plastic bags (freezer zip locks are best), squeeze out the air, then place the plastic bags in duffle bags with shoulder straps or camping backpacks to facilitate transportation. Suitcases are often heavier than canvas bags and many need to be carried by hand. This would be a serious strain for your back and shoulders if you had to carry them a long distance.

Think about your clothing requirements and update clothing sizes for summer and winter. Remember that in winter it is best to dress in layers, so you can remove or add little by little to adjust to your requirements.

Pack the following items according to season:
- 5 to 6 each of your socks, underwear, and brassieres (you do not know when you will be able to wash your clothes)
- Undershirts and/or t-shirts
- Pants and long-sleeved tops
- Sweaters (with hoods if possible, so you can also layer head coverings)
- Sweatshirts and pants, shorts, pyjamas
- Hats, baseball caps, gloves, mittens, scarves
- Winter coats, snow pants, slush pants

- Raincoat (if it is too warm for a winter coat)
- Shoes and slippers
- Sandals or flip-flops (a good idea for showers where many people are using or sharing the facilities—bacteria thrive on wet floors)
- Extra winter boots
- Regular hygiene kit
- Toothpaste and toothbrush
- Birth control, razor, and shaving gel
- Soap, cloth and towel, shower cap
- Feminine hygiene products
- Diapers and wipes, rash cream
- Brush, comb, and shampoo, conditioner
- Face and body cream
- Nail clippers and tweezers
- Eyeglasses, sunglasses
- Contact lenses and solutions
- Jewellery, watch, wallet, purse, keys.

NOTE: Make a list of items that have to be packed at the last minute.

7 | Important Miscellaneous Items

This list starts with the most important items.

Money: coins, bills, travelers checks (if previously purchased), credit cards, ATM cards, bankbooks, check books.

Documents: birth certificates, marriage certificate, passports, driver's license, medical records, hospital cards, medical insurance coverage documents, life insurance, home insurance, last will and testament, power of attorney (all sealed in zip lock bags). If you have too much and must leave some behind, store them in a proper water and fireproof container.

Activities: books, crossword and puzzle books, pencils, pencil sharpener, crayons, paper, coloring books, cards, travel-size popular games, radio, tape player and tapes, CD player and CDs, portable video games, batteries of all necessary sizes, dolls, small stuffed animals, Matchbox sized cars.

Other:
- Cell phone, charger, and extra minutes cards
- Alarm clock
- Reflector blanket and sleeping bags (can be attached to backpacks)
- Camera equipment and spare film
- Thermos, food-heating equipment (covered in Food Kit List, page 15)

- Flashlights, candles, matches, lighter
- Liquid laundry soap
- Mini hairdryer (in case you are in a shelter with electricity; it is not recommended to go to bed with wet hair since a lot of body heat is lost from the head)
- Pot, pan, ladle, spatula
- Cutlery, sharp knife, can opener
- Cup, plate, bowl, scissors
- Dish soap, cloth, scrubber sponge, towel
- Toilet paper, tissue, moist towel wipes
- ABC fire extinguisher
- Power inverter (connects to your car lighter to give you portable electricity).

8 Finances

If there is a major power failure, ATMs and cash registers will not function. If you do not have a stockpile of coins and low denomination bills, you may have problems. Stores may not be able to give you change for your big bills, and therefore you will be paying a lot more for your necessities. Stores may also not accept credit cards during emergencies.

Include coins and low denomination bills in your teenager's emergency supplies freezer bag, since your teenager may become separated from you. They may have to pay for transportation or buy something to eat before they can join up with you.

Types Of Emergencies
And How To Deal With Them

9 | Earthquakes

Earthquakes are one of the most dangerous natural hazards. If you live in, or close to, a region prone to earthquakes, there are steps you can take to protect your family and home. Begin by securing heavy objects, such as bookshelves and top-heavy furniture, by attaching them by the back to the wall. Secure books by adding fashionable but secure rails in front. Secure plants and china by using anti-skid pads or Velcro. Put anti-skid pads under TVs, VCRs, computers and small appliances. Put childproof latches on cupboards to stop the contents from spilling out.

If you have your own hot water tank, make sure it is secured so it will not tip over and break natural gas or electrical lines.

If your home is equipped with natural gas lines, tape or tie the appropriate wrench on or near the pipe to turn off the gas. Don't shut off the gas unless there is a leak, a fire, or you are advised to do so. If the gas is turned off, don't turn it on again. A qualified technician must turn it on.

Be familiar with all the exits in your building.

During an earthquake:

If you are inside a building, stay inside. Do not use the elevator. If you are caught in the elevator, hit every button and get off as soon as you can. Be prepared for the fire alarm and sprinkler system to activate.

Stay away from windows, fireplaces, mirrors, bookcases and file cabinets, shelves, hanging plants, electrical equipment, and appliances. Get under a heavy desk or table and hang on. If it moves, you move with it. If you can't get under something strong, flatten yourself against an interior wall or inside a door frame while standing up and protect your head and neck. If you are in a wheelchair, get as close as you can to an interior wall or inside a door frame that does not have a door that opens both ways, and lock your wheels. Do not get out of your wheelchair.

Keep your personal kit available always.

If you are outside, go to an open area. Move away from buildings, trees, signs, or any structure that could collapse. Stay away from power lines. If you cannot get away from a building, step into a doorway to protect yourself from falling bricks, glass, plaster, and other debris. Do not go into a courtyard, where you can be trapped.

If you are in a crowded store or public place, do not follow the crowd and rush for the exits. Stay clear of windows, displays, hanging lights and find a place to duck under cover and hold on or run to a main wall and stand against it. If you are in a stadium or theatre, stay in your seat and protect your head with your arms. Do not try to leave until the shaking is over. If possible, bend forward from the waist and keep your head below the level of the seat tops.

If you are in a car, pull over to a safer spot on the soft shoulder, out of the path of traffic. You do not know what is going on ahead of you. Try not to block the road. Stay in the car. Avoid getting on bridges or overpasses or going through underpasses. Move away from buildings, poles, and other standing structures that could fall.

After an earthquake:

Listen to your local radio station for information and updates to hear about the best route to safety and shelter locations.

Be ready for aftershocks. Be careful; structural damage can vary from floor to floor.

Never use the elevator.

Wear protective shoes and clothing; there will be a lot of glass and debris that may still come down or be lying around.

If you live near the ocean, stay away from the waterfront; tsunamis or huge ocean waves may hit.

IMPORTANT: PLEASE READ GENERAL
SAFETY PROCEDURES ON PAGE 89

REVIEW

- Prepare your home by securing objects.
- Secure hot water tank and natural gas lines.
- Have your packs ready to grab and run.
- Be familiar with all exits in your building.

10 | Tsunamis

Tsunamis are ocean waves produced by earthquakes, volcanic erup-
tions, meteorites, or underwater landslides. If a major earthquake is
felt, a tsunami could reach the beach within a few minutes, even
before a warning is issued. A strong earthquake lasting 20 seconds or
more near the coast may generate a tsunami. Sometimes the quake
happens in the ocean and you don't feel it. A noticeable rapid rise or
fall in coastal waters is a sign that a tsunami is approaching. A
tsunami can occur during any season of the year and at any time of
the day or night.

As the waves approach the coast they slow down and get bigger.
They can continue for hours with 5- to 90-minute intervals between
each wave. The first wave is usually not the largest. Waves that are 10
to 20 feet (3 to 6 meters) high can be very destructive and cause many
deaths or injuries. They can travel up streams and rivers. Some low-
lying areas could experience severe inland flooding and a collection
of debris for more than 1,000 feet (300 meters) inland from the waves
in streams and rivers.

Preparing for a tsunami:
Plan your evacuation routes to travel uphill 100 feet (30 meters), or at
least 2 miles (3 kilometers) inland. Every step inland or up may make
a difference. You should be able to reach your safe location on foot

within 15 minutes. Take trial runs of your evacuation routes; familiarity may save your life. Be able to follow your routes at night and during bad weather.

Tie down or remove objects from around your home. Tsunami waves can sweep away loose objects. Some may crash into you or your home.

Listen to your local radio station for information and updates. Local emergency management officials can advise you about the best route to safety and likely shelter locations.

Explanation of tsunami warnings:

Information–a message with information about an earthquake that is not expected to generate a tsunami. Usually only one bulletin is issued.

Advisory–an earthquake has occurred in the ocean that might generate a tsunami. Hourly bulletins advising of the situation will be issued. This is a good time to locate all your family members to make sure everyone knows there is a potential threat and remembers the best way to get to higher ground. You should have your emergency packs ready. Gather all your kits and provisions in preparation for an evacuation.

Watch–a tsunami was or may have been generated, but is at least two hours' travel time to the area in watch status. Local officials should prepare the public for possible evacuation if their area is upgraded to a warning. Elderly, sick, or disabled people and children may take longer to evacuate, therefore allow extra time.

Warning–a tsunami has been generated, which could cause damage; people in the area are strongly advised to evacuate.

After a tsunami:

Continue listening to the radio. The tsunami may have damaged roads, bridges, or other places and made them unsafe. Stay out of buildings if water remains around them. The foundations may sink, floors may crack, or walls may collapse. Once you get into a building, watch for loose plaster, drywall, and ceilings that could fall. Open the windows and doors to help dry out the building. Turn off air conditioners and turn on the heating equipment. Shovel mud while it is still moist to give walls and floors an opportunity to dry. Look around for mould.

Look out for fire hazards. Flammable or explosive material may come from upstream. Fire is the most frequent hazard following floods. Check gas, sewage, water lines, and electric appliances for damage or breaks. Report any line breakages and follow safety procedures in the general safety procedure chapter.

Watch out for animals, especially poisonous snakes that may have come into buildings with the water. Tsunami waters flush snakes and animals out of their homes. The snakes and animals may bite because they are frightened. Use a stick to poke through debris.

Check food supplies. Any food that has come in contact with floodwaters may be contaminated and should be thrown out.

It is also dangerous to be in the floodwater because of gasoline, sewage, chemical and other toxins. Do your best to not expose yourself directly to it.

IMPORTANT: PLEASE READ GENERAL
SAFETY PROCEDURES ON PAGE 89

IMPORTANT: PLEASE READ THE SECTION
ABOUT MOLD ON PAGE 91

REVIEW

- Prepare your home surroundings by securing loose items.
- Plan escape routes.
- As soon as a tsunami advisory is issued, contact family members to advise them of the threat.
- Gather your kits and provisions in preparation for evacuation.
- Listen constantly to the radio or TV.
- Monitor escape routes for road blocks or trouble.
- After the tsunami, inspect building foundations before entering, then check gas, sewage, water lines, and electric appliances for damage or breaks.
- Check food supplies for damage.

11 | Floods

Floods can occur in any region, in the countryside or in cities, at virtually any time of the year.

If you live in a flood-prone area, remove all chemicals, explosive cans, and hazardous products from the basement, as they can float, hit something and break open. Seal all hazardous products such as weed killers or insecticides. Move as much of your precious or special items to the highest level possible.

Familiarize your family members with water safety rules and make sure they know how to swim. If you live near a body of water see that everyone has a life jacket. Make sure everyone knows the boating rules and is aware of the danger of electrocution in the water.

Preparing for a flood:

When authorities have advised you that flooding is imminent, take the following precautions.

Bring in all moveable lawn and yard accessories. Remove toilet bowl water by turning the valve off, then flush. Plug basement sewer drains. If you have a sump-pump, check to see if it is working. Have sandbags ready to use.

Contact the municipality for directions. If you must evacuate, turn off the power and gas and follow the routes specified by officials. Listen constantly to the radio or TV. Don't take shortcuts. They could lead you to a blocked or dangerous area. Follow your evacuation plan. Make sure to take your emergency food and other kits with you.

Never try to wade through a flooded area. Currents can knock you over or sweep you away. Debris both under and on top of the water can also knock you down. Watch out for downed power lines.

Returning home after a flood:

Before entering a flooded building, check for foundation damage and make sure all porch roofs and overhangs are supported. Wear protective clothing (waterproof boots, long pants, long sleeved shirt, gloves, safety goggles, etc.) Stay out of rooms where ceilings are sagging from retained water. If you have a basement full of water, drain the water in stages, about a third of the volume per day. Draining water too quickly can structurally damage your home. Disinfect the stagnant water every three days if the flood is severe and you are back living in your home. For an average size home, pour approximately a half a gallon (2 liters) of liquid bleach directly into the floodwater if it is a few feet deep. Ventilate wet areas. Turn off air conditioning for accelerated drying in summer. In winter, alternate cycles of opening windows and turning on heating equipment.

IMPORTANT: PLEASE READ GENERAL
SAFETY PROCEDURES ON PAGE 89

READ THE SECTION ON MOLD ON PAGE 91

R E V I E W

- If there is boating, ensure everyone has a life jacket, knows how to swim, and knows boating safety rules.
- When advised of flood warning:
  - Prepare your home inside, and secure your home outside.
  - Listen constantly to the radio or TV.
  - Follow official escape routes.
- Upon return:
  - Wear protective clothing.
  - Check for foundation damage and home structure integrity.

42

12 | Hurricanes

Hurricanes have wind speeds from 73 miles (117 km) to over 100 miles (160 km) per hour, usually cover a large area and bring thunder, lightning, and heavy rain. Tornadoes can often follow a hurricane. The destruction can get worse once the eye, which is deceptively calm, passes over and the winds blow from the opposite direction as they swirl in a circle. Trees, shrubs, buildings, and other objects damaged by the first winds can be broken or destroyed by the second winds and be thrown around.

Preparing for a hurricane:

Prepare your evacuation plans so you are heading as far inland and away from the coast as possible since the storm loses force as it hits land. If this is not possible due to lack of transportation, identify ahead of time where you should go if you are told to evacuate. Make sure your emergency kits and provisions are prepared. Depending on the force of the hurricane or the landfall projection, you may not be evacuated.

However, it would be wise to do as much as possible to protect your home. Install hurricane shutters or purchase pre-cut 1-inch outdoor plywood boards for each window of your home. Install anchors for the plywood and pre-drill holes in it so you can put them up quickly. Tape does not prevent windows from breaking, so it is not recommended.

Make trees more wind resistant by removing dead or damaged limbs, then removing some branches so that wind can blow through. Prepare to bring inside any lawn furniture, outdoor decorations or ornaments, trash cans, hanging plants, and anything else that can be picked up by the wind.

At first notification of a hurricane threat, contact your family and advise them of your plans.

If you are not evacuated, remain indoors and always stay away from windows. Stay in the center of your home, in a closet or bathroom without windows. Listen constantly to your portable radio for updates and instructions.

If you are asked to evacuate, do not delay. Remember there are thousands of other people doing the same and roads may be clogged or closed. Stay away from floodwaters. If you come upon a flooded road, turn around and go another way. If you are caught on a flooded road and waters are rising rapidly around you, get out of the car and climb to higher ground.

Explanation of hurricane warnings:

Hurricane watch–hurricane conditions are possible in the specified area of the watch, usually within 36 hours.

Hurricane warning–hurricane conditions are expected in the specified area of the warning, usually within 24 hours.

IMPORTANT: PLEASE READ GENERAL
SAFETY PROCEDURES ON PAGE 89

R E V I E W

- Prepare evacuation plans.
- Prepare kits to grab and run.
- Prepare your home to prevent outside damage.
- At first notification of a hurricane threat, contact your family and advise them of your plans.
- Listen constantly to the radio or TV as the hurricane nears for updates and instructions.
- Evacuate immediately upon advice from officials.
- Watch out for floodwaters.

13 | Tornadoes

Tornadoes are rotating columns of high wind with speeds up to 310 miles (500 km) per hour or more that destroy anything in their path. They often follow a severe thunderstorm and are difficult to predict. They can travel slowly or quickly and can leave a wide path of destruction. Tornadoes are also small, touching down and leaving a skipping damage path. They can uproot trees, flip cars, and demolish houses. There may be lightning, hail and fast, swirling winds up to a mile from the tornado center.

During a tornado:

Keep tuned to your local TV and radio stations for updated storm information. Be familiar with safety tips in case you are caught away from your safe area.

If you are home, go to the basement with all your kits or take shelter in a small interior ground floor room such as a bathroom, closet, or hallway. If that is not possible, take shelter under a heavy table or desk. You can also pull a mattress on top of you to cushion the blow of falling objects. In all cases, stay away from windows, outside walls, and doors.

If you are at the office or in a high rise building, take shelter in an inner hallway or room. Try to get to the basement or ground floor. Do not use the elevator and stay away from windows.

Avoid buildings such as gymnasiums, churches, arenas, shopping centers, and auditoriums with high roofs. These roofs do not have supports in the middle and may be blown off or collapse if a tornado hits close to them.

If you are caught by surprise outside, or in a mobile home, and no shelter is available, lie down in a ditch or ravine away from the mobile home. If a tornado seems to be standing still, then it is either traveling away from you or heading right for you. If you are driving and spot a tornado in the distance, try to get to nearby shelter. Turn and drive away from the tornado. Do not try to beat it to another road or cut-off. If the tornado is close by, get out of your car and take cover in a low lying area or even under an underpass on a freeway. Crawl right up the bank to the joints of the overpass. No matter what your situation is, get as close to the ground as possible and protect your head. Stay there until all the winds carrying debris have passed.

After the tornado:

If your home or family is affected by a tornado, you should monitor local media reports for advice and to find out where assistance is available.

IMPORTANT: PLEASE READ GENERAL
SAFETY PROCEDURES ON PAGE 89

R E V I E W

- Be familiar with safety tips in case you are caught away from your safe area.
- Go to your safe area.

14 | Hail, Thunder, and Lightning Storms

For all storms, seek appropriate shelter. Stay sheltered for at least 30 minutes after the last piece of hail, lightening strike or thunderclap. Hail forms in thunderstorms, can be extremely dangerous and can cause extensive damage in only a few minutes. It can fall at speeds of up to 60 miles (100 km) per hour. It often appears near the area in a thunderstorm where tornadoes may form. Once large hail starts to fall, it is safer to assume that a tornado could be nearby and to seek appropriate shelter.

Thunder may have a sharp cracking sound when lightning is close by. To judge how close lightning is, count the seconds between the flash and the thunderclap. If you count less than 30 seconds, the storm is less than 6 miles (10 km) away. If you count less than 5 seconds, take shelter, preferably in a house, automobile (not with a convertible top) or in a low-lying area.

Lightning strikes can carry up to 100 million volts of electricity and leap from cloud to cloud or cloud to ground and vice versa. Lightning strikes higher ground and the tallest objects around. Trees and metal objects are especially good conductors of electricity. Precautions should be taken even if the thunderstorm is not directly overhead.

Outside during a storm:

If you are caught outside, keep away from tall objects, such as trees. Never be the tallest object when you are out in the open. Seek shelter in low-lying areas such as valleys, ditches, and depressions. Crouch down to become as small as possible and protect your head and neck. Do not lie down; you want to be the smallest compact object around.

Stay away from water. If you are boating or swimming and a storm threatens, get to land as quickly as possible. Stay away from objects that conduct electricity such as tractors, golf carts, golf clubs, metal fences, motorcycles, lawnmowers, and bicycles. Do not hold golf clubs, an umbrella or fishing rods. Take off shoes with metal cleats or eyelets.

You are safe in a car during a lightning storm, but don't park near or under trees or other tall objects that may topple over during a storm. Watch out for downed power lines that may be touching your car. If you must get out of the car for safety reasons, jump away from the vehicle so you don't get a shock if a power line is on the car or touching water. Do not touch the vehicle and the ground at the same time. In a forest, seek shelter in a low-lying area under a thick growth of small trees or bushes.

If you are caught in a flat field far from shelter and you feel your hair stand on end, lightning may be about to hit you. Crouch down on the ground immediately. Don't lie flat. If you are in a group in the open, spread out, keeping people several yards apart.

Inside during a storm:

Don't go outside during a storm unless absolutely necessary. Before the storm hits, disconnect electrical appliances including radios, air conditioners, computers, and television sets. Power surges can overload your wiring, burn your appliances or even start a fire. Do not touch appliances during the storm. Don't use the telephones; use cell phones and battery operated appliances only. Keep away from doors, windows, fireplaces, and anything that could conduct electricity, such as radiators, stoves, sinks, and metal pipes. Keep as many walls as possible between you and the outside. Close blinds and shades over the windows to stop glass from flying freely into the house if windows break. In addition, avoid baths, showers, or running water, as water conducts electricity.

People who have been struck by lightening do not carry an electrical charge and can be safely handled. They may have burns or be in shock and should receive medical attention immediately. If the person is not breathing, give CPR (cardiopulmonary resuscitation) right away.

IMPORTANT: PLEASE READ GENERAL
SAFETY PROCEDURES ON PAGE 89

IMPORTANT: PLEASE READ THE FIRST AID
CHAPTER FOR EMERGENCIES ON PAGE 83

R E V I E W

- For all storms, seek appropriate shelter.
- Stay sheltered for at least 30 minutes after the last piece of hail, lightning strike, or thunderclap occurs.
- If you are caught outside, crouch down and protect your head and neck.
- Never be the tallest object when you are out in the open.
- Seek some form of shelter.
- If you are in your car, watch out for downed power lines or any other items that may topple over on you.
- If you are at home, disconnect electrical appliances. Stay away from any metal that can conduct electricity. Do not use the phone. Stay away from windows and close the blinds.
- Stay away from water.
- Victims of a lightning strike can be handled. Administer first aid and call for an ambulance as soon as possible.

15 | Landslides

Landslides are a serious hazard and are common. Some move slowly and cause damage gradually. Others move so fast, up to 35 miles (56 km) an hour or more, that they can destroy property and take lives suddenly and unexpectedly.

Landslides usually happen during or after earthquakes, heavy rainfall or rapid snow melt. They usually start on steep hillsides or areas that have been cut open for roads. Areas burned by forest and brush fires have no trees to hold the land in place. As they descend landslides pick up debris and can move homes right off their foundations. They contain mud, from watery to thick and rocky, that can carry away such large items as boulders, trees, and cars.

Preparing for a landslide:

Learn about landslide risk in your area. Prepare your emergency kits. Be alert for patterns of storm-water drainage on slopes near your home. Check hillsides around your home for any signs of land movement such as small landslides or tilting trees.

If you are at risk from landslides, develop an emergency and evacuation plan. Inform neighbors, since they may not be aware of the hazards. Advising them of a threat may help save lives. Help plan for neighbors who may need assistance to evacuate.

Have a professional install flexible pipe fittings to avoid gas or water leaks, especially on outdoor pipes. Flexible fittings are less likely to break.

During a storm:

During intense storms, stay alert and awake. Fatalities occur when people are sleeping. Listen to the portable radio for warnings of intense rainfall. Listen for any unusual sounds such as trees cracking or boulders knocking together. A small flow or mud slide may come just before a larger one. Sometimes there is no warning.

If you are near a stream keep an eye on the water flow and for any change from clear to muddy water. Changes may indicate landslide activity upstream, so be prepared to move quickly.

If you are in an area that has had landslides, consider leaving if it is safe to do so. Remember that driving during an intense storm can be hazardous. Watch the road for collapsed pavement, mud and fallen rocks.

If you remain at home, move to a second story if possible. If staying out of the path of a landslide is not possible, curl into a tight ball and protect your head. Watch for flooding, which may occur after a landslide or debris flow. Floods sometimes follow landslides and debris flows because they may both be started by the same event.

Report any activity to emergency services. Assist those in need if possible without endangering yourself, as you must move quickly.

IMPORTANT: PLEASE READ GENERAL
SAFETY PROCEDURES ON PAGE 89

REVIEW

- Learn about the landscape in your area.
- Develop an emergency and evacuation plan.
- Prepare your kits.
- Prepare your water and gas pipes against possible breakage.
- During intense storms, stay alert and awake. Listen for unusual sounds.
- Listen to the radio or TV for warnings of expected intense rainfall.

- Be alert for outdoor changes.
- Inform neighbors of impending landslide possibility.
- Assist those in need if possible without endangering yourself, as you must move quickly.
- If you remain at home or are caught in a landslide, move to a higher level in the home. If escape is not possible, curl into a tight ball and protect your head.
- Watch for post-landslide flooding.

| 16 | **Heat Waves** |

Heat waves are several days of abnormally hot weather. They can cause serious problems. Care for the elderly and young children is important, as their bodies do not regulate heat efficiently. When our bodies get too hot, the heart pumps more blood to the skin and this causes the skin to sweat. As sweat evaporates, it cools down the body. But if you sweat too much you lose too much fluid and dehydrate. Learn the signs and treatments for heat related attacks.

During a heat wave:

Slow down. Your body can't do its best in high temperatures. Stay out of the sun; you don't want a sunburn. If you must be in the sun, use sunscreens. Cool your body with showers or baths, fans or air conditioners. Drink plenty of water to keep your body from drying out. Maintain salt levels in your body. If you are on a salt-free diet, check with your doctor. Avoid high-protein foods as they increase your body's water loss and heat production (e.g., red meat, beans, peas, nuts, fish, eggs, poultry, and milk products). Dress in lightweight and light-colored clothing.

IMPORTANT: PLEASE READ FIRST AID SECTION
FOR HEAT RELATED ILLNESSES ON PAGE 84–86

R E V I E W

- Avoid overheating, stay out of the sun, dress appropriately and slow down.
- Drink plenty of water. Maintain salt levels in your body.
- Avoid high protein foods.
- Learn the signs and treatments for heat-related attacks.

17 | Blizzards and Ice Storms

Winter weather conditions can become severe or dangerous quickly, sometimes with little or no warning. High wind chills, heavy snowfall, freezing rain, blizzards, and bitterly cold temperatures all pose a hazard to those venturing outside or traveling.

Blizzards are severe winter storms with heavy blowing snow, winds of 25 miles (40 km) or more per hour. Visibility can be reduced to a few yards (meters). They must last for four hours or more to be officially classified as blizzards. They can last for days. Squalls are short duration blizzards that can have the same effect.

Preparing for a blizzard:

Check the weather forecast before traveling or pursuing outdoor activities. Pay particular attention to the wind chill factor; it can create dangerous conditions.

Winterize your home and vehicle before the cold weather season. Prepare the emergency packs for your home and vehicle and ensure your home heating system is in good working order. Insulate your home to avoid cold air leaks and minimize heating costs. Weather stripping and caulking around doors and plastic shrink packs around your windows are cost-efficient and easy to install. Service snow removal equipment and have rock salt on hand to melt ice on walkways and kitty litter to generate temporary traction.

During a blizzard:

Make sure your kits are prepared. Do not go outside; you may lose your way in the blinding snow. If you must go out, firmly tie one end of a long rope to yourself and your house to provide a way back, and go only for a short distance. Keep a tight grip on the rope.

Be prepared for power failures. If you lose power, turn the thermostats down to minimum and turn off all appliances, electronic equipment, and tools to prevent injury, damage to equipment, and fire in case the power surges when it returns. Power can also be restored more easily when the system is not overloaded. If you use candles, make sure to use the proper candleholders. Never leave lit candles unattended.

Be alert for fire hazards due to overheated stoves, fireplaces, heaters, or furnaces. Don't use charcoal or gas barbecues, camping heating equipment, or home generators indoors. Use heating equipment that is certified for indoor use only. Home generators are handy for back-up electricity in case of an outage, but there are hazards to be aware of, like the oil running out, keeping it clear of snow drifts, etc. You must investigate thoroughly before purchasing this item. It is very costly.

Keep babies warm. Put on a diaper, undershirt, sleeper, sweater, blanket, and or light bunting. Keep a light hat on your baby's head (70 per cent of body heat is lost through the head). Keep your baby close to you; your body heat will keep it warm and the closeness will reduce anxiety. Your baby can sleep with you in your bed to stay warm, but be careful not to let the child roll out of the bed, or you roll on top of it. Babies should never sleep or be placed on a waterbed, for risk of suffocation. Do not fully bathe your baby. Keep its face and bottom clean by using a washcloth or baby wipes.

In your home let faucets drip a little to avoid freezing. Know how to shut off water valves. To keep pipes from freezing, wrap them in insulation or layers of old newspapers. Cover the newspapers with plastic to keep out moisture. If the pipes freeze, remove any insulation or layers of newspapers and wrap pipes in rags. Completely open all faucets and pour hot water over the pipes, starting where they were most exposed to the cold, or where the cold was most likely to penetrate.

Conserve fuel. Lower the thermostat to 65 degrees Fahrenheit (18 degrees Celsius) during the day and 55 degrees Fahrenheit (13 degrees Celsius) at night. Close off unused rooms. Keep warm by dressing in layers. Pay particular attention to your hands and feet. Move them often. Elderly persons should be especially careful, as their bodies do not adjust as easily to temperature change. Keep active and busy; moving around will help your circulation.

Bundle up more at night as your body temperature drops while asleep. Use extra blankets. Family members could sleep in the same bed to keep warm. Hot water bottles or other heating devices can be used to warm the bed, but remove them before sleeping to prevent burns.

If you must go outside, dress to suit the weather. Thin layers of loose fitting clothing will trap body heat while aiding air circulation. Layers can be put on or taken off easily to adjust your body temperature. Outer clothing should be hooded, tightly woven, and repel water. Mittens provide more warmth than gloves because your fingers rubbing together help keep them warm. Because most body heat is lost through the head, it is important to wear a hat. If it is extremely cold, cover your nose and mouth with a scarf to protect your lungs from the cold air. Keep dry. Change wet clothing often to prevent the loss of body heat.

Pace your outdoor activities, and be alert for signs of frostbite or dehydration. Watch for signs of hypothermia, which is particularly threatening to the very young and elderly.

IMPORTANT: PLEASE READ FIRST AID
SECTION FOR FROSTBITE, DEHYDRATION
AND HYPOTHERMIA ON PAGE 86–87

Being stranded in your car during a blizzard can be traumatic and dangerous even after you have pulled out of traffic and parked on the side of the road. Make sure that snow does not pile up and block your tail pipe, as the exhaust from running your engine cannot exit the car and will back up inside and suffocate you.

You may be seen as a vulnerable target in your car. Purchase a Call Police sign, available at most retail outlets. The sign is durable and reflective and can be seen in both directions at night and in any kind of weather. It allows you to ask for assistance without leaving your vehicle. Pull the vehicle completely off the road, turn your emergency flashers on, roll the driver's side window down, and hook the sign on. Then roll the window back up, lock all doors, and remain in the vehicle. Open a back or side window half an inch or 1 cm for ventilation.

If someone other than the police approaches your vehicle, DO NOT open your window or unlock the doors. Open a back window a crack to ask the motorist to please call the police using a cell phone, or at the nearest service station or store. If you see a motorist stranded on the side of the road, DO NOT stop. Instead, call 911. Note the location of the vehicle by using street signs or highway markers, and advise the police.

Use caution when shoveling after a storm. Stretch before you go out; it warms up your body. Take frequent breaks. It is extremely hard work and can even bring on a heart attack or make other medical conditions worse.

During an ice storm:

Ice storms often have a worse effect than blizzards. More provisions must be organized and planned for than for blizzards, as snow can be removed from the roads more easily than thick ice.

Drivers should not go out during ice storms, as there are unforeseen dangers that can create problems in a split second. People are often preoccupied and overwhelmed by the state of emergency and may not drive safely when they are distracted. Tree branches and power lines can snap and fall without warning, and roads may have black ice patches. Vehicles may not be able to brake at stop signs or traffic lights, which may be malfunctioning. Other drivers may just pass through without stopping. A four-way intersection becomes a four-way stop when the lights are not working. The first vehicle to arrive and stop has the right of way. If two or more vehicles stop at the same time, the vehicle on the right has the right of way.

IMPORTANT: PLEASE READ GENERAL
SAFETY PROCEDURES ON PAGE 89

R E V I E W

- Winterize your home before the cold weather arrives.
- Prepare your emergency kits.
- If you absolutely must go out in inclement weather:
    - Dress in layers to suit the weather.
    - Watch for hypothermia and frostbite.
    - Pay attention to the wind chill factor.
    - Tie a rope from your home to yourself or destination and hold on constantly as you walk.
- Stretch before going out to do strenuous activities such as shoveling. Pace yourself.
- Be prepared for power failures.
- Be alert for fires due to overheated equipment.
- If you are stranded on the road, pull over as far as possible and turn your emergency flashers on. Make sure your tailpipe does not get filled up with snow and block the exhaust. Keep one window open half an inch or 1 cm for ventilation.

18 | Fire

Almost all fires are small at first, and most can be easily extinguished if the proper type and amount of extinguishing agent is promptly applied. Portable fire extinguishers are designed for this purpose, but their successful use depends on the extinguisher being properly located and in good working order. It also must be the proper type for the fire. Most often, the ABC type will be the best. It is designed to extinguish ordinary combustible materials such as wood, cloth, paper, many plastics, flammable or combustible liquids such as petroleum products and greases, and electrical fires.

For all fires:
If you smell gas, smoke or see a fire, you must act without delay. Pull the fire alarm or scream "Fire". If it is a small fire and you are able to do so safely, get an extinguisher. If not, close all doors possible, especially to the room that is on fire. Leave the building immediately. Never use an elevator. Call 911 or the fire department immediately. Help those who may be physically impaired, the elderly and children. Once you are out, stay out. Don't let anyone go inside for any reason. Report any persons still in the building or missing. Go upwind and as far away as possible because there could be an explosion, drifting smoke or the fire could spread to the next building. Do not go into a courtyard where you may become trapped.

Always aim the nozzle of an extinguisher at the base of the fire, using sweeping motions from side to side. DO NOT attempt to stand over the fire and direct the contents of the extinguisher down onto the flames. That will cause the flames to spread outward.

When a fire occurs, seconds count. Everyone must know how to react swiftly and smartly to stop or contain the fire, alert others to the emergency, or to escape and sound warnings.

Preventing a home fire:

Make your home fire-safe. Smoke alarms save lives. Install a smoke alarm outside each sleeping area and on each additional level of your home. If people sleep with the doors closed, install a smoke alarm inside sleeping areas too. Use the test button to check each smoke alarm once a month. When necessary, replace batteries immediately. Replace all batteries at least once a year. Vacuum away cobwebs and dust from your smoke alarm monthly. Smoke alarms become less sensitive over time. Replace your alarms every 10 years maximum. Have one or more working fire extinguishers in your home and one in your car. In your home they should be placed in open accessible places like the kitchen.

Have a ladder that can reach the roof in case you need to get someone down from an upper level. In addition, consider installing an automatic fire sprinkler system in your home. Inspect all lights, extension cords, wires and plugs on your appliances. Replace them if they are cracked, frayed or the old type that are not up to code.

Do not overload extension cords or outlets, or run cords under rugs. Do not tamper with fuse boxes.

Do not store flammable items near rags or other flammable goods.

Keep portable heaters and space heaters at least 3 feet (1 meter) away from anything that can burn. Never leave heaters on when you leave the room.

When you barbeque, keep a clean area around you, place a screen over the grill and use non-flammable material with one-quarter inch mesh to help prevent flames and sparks from coming through and lighting you or your surroundings on fire.

Plan your escape routes. Determine at least two ways to escape from every room of your home. Consider escape ladders for sleeping areas on the second or third floor. Learn how to use them and store them near or attached to the window. Select a location outside your home where everyone is to meet after escaping. Practise your escape plan at least twice a year.

During a home fire:

If you see smoke or fire in your first escape route, use your second way out. If you must exit through smoke, crawl low to the ground, keeping your head about 12–24 inches (30–60 cm) above the floor and under the smoke as you exit. During a fire, smoke and poisonous gases rise with the heat. The air is cleaner near the floor.

If you are trying to escape through a closed door, kneel or crouch at the door, reach up as high as you can with the back of your hand or arm (it is more sensitive), and touch the door, the knob, and the crack between the door and its frame. If you feel any warmth at all, use another escape route. If the door feels cool, open it with caution. Put your shoulder against the door and open it slowly. Be prepared to slam it shut if there is smoke or flames on the other side. If smoke, heat, or flames block your exit routes, stay in the room with the door closed. Signal for help by waving a brightly colored cloth out the window. If there is a telephone in the room, call the fire department and tell them where you are.

If you live in a high-rise apartment, familiarize yourself with the locations of all fire alarm pull stations, fire extinguishers or fire hoses. Follow the previous directions for all fires.

Electrical Fire:

If you see smoke or fire and then smell a strange odor coming from the wires, appliance or electric motor, turn off the appliance, pull out the cord and turn off the main switch at the circuit breaker or fuse box.

DO NOT throw water on an electrical fire; you could be electrocuted. Use a class C fire extinguisher. If no appropriate fire extinguisher is available, you can also use baking soda to extinguish an electrical fire. Call 911 or the fire department right away.

Wildfire:

More and more people are making their homes in woodland settings, in or near forests, rural areas, or remote mountain sites. Wildfires often begin unnoticed. They spread quickly. Reduce your risk by preparing now. Begin by reporting hazardous conditions that could cause wildfire. Teach children about fire safety, and keep matches and lighters out of their reach.

Plan several escape routes away from your home, by car, and by foot. Make plans and a list of your neighbors' skills, such as medical or technical abilities. Consider how you could help neighbors who have special needs, such as elderly or disabled persons. Make plans to take care of children who may be on their own if their parents cannot get home.

Inspect chimneys at least twice a year. Clean them at least once a year. Keep the dampers in good working order. Equip chimneys and stovepipes with a spark arrester. Use mesh screen beneath porches, decks, floor areas, and the home itself to block leaves and other flammable material from blowing underneath. Consider installing protective shutters or heavy fire-resistant drapes.

Rake leaves, dead limbs, and twigs. Remove leaves and rubbish from under structures. Create a 15-foot (4.5-meter) space between tree crowns and remove limbs within 15 feet of the ground. Remove dead branches that extend over the roof. Prune tree branches and shrubs within 15 feet (4.5 meters) of a stovepipe or chimney outlet. If there's no strong wind the extra space may keep the flames from jumping to your building. Ask the power company to clear branches from the power lines. Removing vines from the walls of the home will also help to eliminate danger. Mow grass regularly. Clear a 10-foot (3-meter) area around propane tanks and the barbecue.

Regularly dispose of newspapers and rubbish. Follow local burning regulations if you dispose of your own garbage. Store gasoline, oily rags and other flammable materials in approved safety cans. Place cans in a safe location away from buildings.

Stack firewood at least 100 feet (30 meters) away and uphill from your home. Clear away combustible material within 20 feet (6 meters) of your home. Use only UL-approved wood-burning devices. Place stove, fireplace and grill ashes in a metal bucket, soak in water for two days, and then bury the cold ashes in wet soil.

Identify and maintain an adequate outside water source such as a small pond, cistern, well, swimming pool, or hydrant. Have a garden hose that is long enough to reach any area of the home, roof, and other structures on your property. Install freeze-proof exterior water outlets on at least two sides of the home and near other structures on the property. Install additional outlets at least 50 feet (15 meters) from the home if you have a large lot. Consider obtaining a portable gasoline-powered water pump in case the electrical power is cut off and you are in a remote area.

If there is a warning that a wildfire is threatening your area, listen to your battery operated radio for reports and evacuation information. Back your car into the garage or park it in an open space facing the direction of escape. Shut doors and roll up windows. Leave the key in the ignition. Close garage windows and doors but leave them unlocked. Disconnect automatic garage door openers in case you lose power and cannot open the door quickly. Confine pets to one room in case you have to evacuate; they will be easy to locate and pick up.

Wear protective clothing, sturdy shoes, cotton or woolen clothing, long pants and shirts, gloves, and carry a handkerchief to protect your face. Take your supply kits. Do not delay to gather and save items. Choose a route away from the fire hazards. Watch for changes in the speed and direction of fire and smoke.

IMPORTANT: PLEASE READ FIRST AID
SECTION FOR BURNS ON PAGE 84

IMPORTANT: PLEASE READ GENERAL
SAFETY PROCEDURES ON PAGE 89

REVIEW

- Install fire alarms and test them monthly.
- Plan escape routes and practise them twice a year.
- Check all wires and plugs for fraying or cracking.
- Make sure there is an ABC fire extinguisher at home, at the office, and in the car.

- Have baking soda handy for grease or electrical fires if there is no fire extinguisher handy.
- Secure your home and clean up the surrounding outdoor area.
- Listen to the radio or TV if a fire is threatening your area.
- Prepare your kits to evacuate.
- Wear protective clothing.
- In case of fire, act quickly. Do not delay to gather and save items.
- Sound the alarm.
- If caught in a fire, stay low and check doors before opening them. Stay low to the ground to avoid smoke and chemicals.
- Once out, stay out. Stay at least 300 feet (90 meters) upwind of the fire.

Chemical and Biological Emergencies

Chemicals can be poisonous or have a harmful effect on your health. Some chemicals, which are safe and even helpful in small amounts, can be harmful in larger quantities or under certain conditions.

Remember, you may be exposed to chemical or biological hazards even though you may not be able to see or smell anything unusual. You can be exposed in three ways: breathing, swallowing something contaminated (food, water or medication), or touching the hazard directly, even touching clothing or things that have come in contact with the hazard.

Prevention:

Chemical accidents can be prevented. Most people think of chemicals as part of manufacturing procedures. Chemicals are in fact found everywhere in our homes. It is important to childproof your home. Affix latches or locks on all cabinets that contain medication, cleaning detergents, and other harmful agents.

Never mix one agent with another. Some common household chemical products produce toxic gases (for example, ammonia and bleach). Always follow directions for use. Some products must be used in open spaces to avoid inhalation, or with protective clothing and glasses. Never smoke while using household chemicals. Don't use hair spray, cleaning solutions, paint products, or pesticides near the open flame of an appliance, pilot light, lighted candles, fireplace, wood burning stove, etc. Although you may not be able to see or smell the vapor, particles in the air could catch fire or explode.

Always store agents according to directions and never near food. Never dispose of a chemical in the garbage. You can harm yourself, members of your family, accidentally contaminate your local water supply, or harm your neighbors. Call your local authorities and ask them for proper disposal procedures, such as in a hazardous waste collection facility or recycling center.

Chemical poisoning:

There are several symptoms of chemical poisoning, whether exposure is due to breathing, swallowing, or touching. These include: difficulty breathing, changes in skin color, headache or blurred vision, dizziness, irritated eyes/skin/throat, unusual behavior, clumsiness or lack of coordination, stomach cramps or diarrhoea.

Teach your children what the skull and crossbones and the explosive signs mean. Teach your children to never take medicine or pills without you giving it to them. If your child eats or drinks a non-food substance, find the container and bring both the child and container to the phone. Call the poison control center or 911. Follow their instructions carefully. Often the first aid advice found on containers may not be appropriate. Do not give anything by mouth until you have been advised to do so by medical professionals.

If you think you have been exposed, you see or smell something that you think might be dangerous or find someone who has been overcome with toxic vapors, your first job is to make sure that you don't become a victim. If you remain in a dangerous area and become injured or unconscious, you cannot help yourself or any other victims. Call or send someone else to call 911 right away. Describe what has happened, how many people are involved, and what is being done to help. Stay on the phone until the operator tells you to hang up.

MAJOR CHEMICAL OR BIOLOGICAL EMERGENCIES

A major chemical emergency is an accident that releases a hazardous amount of a chemical into the environment. Accidents can happen underground, on railroad tracks or highways, and at manufacturing plants. These accidents sometimes result in a fire or explosion, but many times you cannot see or smell anything. If this should occur, you will be notified by the authorities. To get your attention, a siren may be sounded, you may be called by telephone, or emergency personnel may drive by and give instructions over a loudspeaker. Officials may even come to your door. The danger will also be announced over the radio and TV.

You must strictly follow all instructions as your life could depend on it. You will be told the type of health hazard, the affected area, how to protect yourself, whether to stay in your home or to evacuate, where the evacuation routes are, and the type and location of medical facilities.

Shelter in place:

Usually one of the first instructions you will be given is to shelter in place. In this case, take your children and pets indoors immediately. You can do this while protecting your breathing by covering your mouth and nose with a damp cloth.

Close all windows in your home. Turn off all fans, heating, and air conditioning systems. Close the fireplace damper. Go to an above-ground room with the fewest windows and doors, NOT the basement because chemicals are usually heavier than air. Take your emergency supply kits with you. Wet some towels and jam them in the crack under the doors. Tape around door frames, windows, exhaust fans, or vents. Use plastic garbage bags to cover windows, outlets, and heat registers.

If you live in an area where you are close to a chemical plant, you should have duct tape and large plastic sheets or rolls in your emergency supply kit. If you are told there is a danger of explosion, close the window shades, blinds, or curtains to block glass from flying around the room. Stay in the room and listen to your radio until you are told all is safe, or until you are told to evacuate.

Evacuate (very low probability):

Take your emergency supplies kits, shut off all vents, and turn off appliances except the fridge and freezer. Close and lock your windows and doors. Move quickly and calmly. If you need a ride, ask a neighbor. If no neighbor is available to help you, listen to the emergency broadcast station for further instructions.

Take only one car to the evacuation site. Close its windows and air vents, and turn off the heater or air conditioner. Don't take short cuts as they may put you in the path of danger. Follow the exact route you are told to take. If your children are at school, DO NOT go to the school. You should already know the procedures the school has in place. Staffs are trained to handle emergencies. Do not phone them. You could tie up a phone line that is needed for emergency communications. For further information, listen to local emergency radio and TV stations to learn when and where you can pick up your children.

For chemical and biological warfare, there are no easy fixes. Some chemical agents act within minutes. By the time an attack has been identified, it may be too late to save the first victims. By the time the type is discovered and treatment is determined and implemented, victims may perish. Unless there is an early warning of the impending attack, there is really no way to protect the general population.

The moment you are alerted, you can somewhat protect yourself by putting on a dust mask (obtainable at any pharmacy), or placing a rag or handkerchief dipped in a solution of 1 spoonful of baking soda per cup of water over your nose and mouth. To protect your eyes, wear swimming goggles. Wear rubber gloves or plastic bags held in place with elastic bands over your hands and wrists. Wear a waterproof poncho or raincoat and hood to protect your body and head from direct contact. Waterproof boots are also required.

If you are at home, keep all your emergency kits with you. Stay in a closed room without windows. If this is not possible, close all widows, air vents, and doors. Tape all edges; lay a wet towel along the bottom of the door to block air. Tape the keyholes and eyeholes on doors. Eat canned food or food and drinks sealed in glass containers. Listen to the TV or radio and do not leave the room until an all clear is given.

IMPORTANT: PLEASE READ GENERAL
SAFETY PROCEDURES ON PAGE 89

IMPORTANT: PLEASE READ THE FIRST AID
SECTION ON PAGE 83

REVIEW

- Childproof your home. Put latches or locks on all cabinets that have medicines or cleaning agents.
- Teach your children to understand and beware of the crossbones and explosive signs and never to take medicines or pills without your permission.
- Always follow directions for use on the labels for medicine and cleaning or chemical products.
- Never dispose of chemicals in the garbage.
- In case of chemical emergency:
  - Listen to the radio or TV.
  - Bring your children and pets indoors immediately.
  - Secure your home air quality immediately.
  - Follow the exact escape routes advised. Close car vents, heater, air conditioner, and windows in the car.
  - For swallowed, touched, or breathed-in chemicals, call 911 or poison control immediately.
  - Shelter or secure yourself and call 911 immediately if you suspect something is wrong. Wait for instructions from authorities.
- For chemical and biological warfare:
  - The moment you are alerted, wear protective clothing.
  - Secure your home, listen to the TV or radio, and do not leave your secure area.

20 | Terrorism

Terrorists look for crowded areas where they can blend in before or after an attack, such as international airports, large cities, major international event sites, resorts, and high-profile landmarks.

You can prepare for the unexpected and reduce the stress that you may feel now, and later, should another emergency arise. Finding out what can happen is the first step. Discuss it with your family and prepare your plans and emergency kits as previously outlined.

A daily routine tends to make you complacent. Take a second look around your area. You need to be aware of your surroundings because terrorists don't give warning.

Take precautions when traveling. Look around you. Be aware of strange or unusual behavior. Do not accept packages from strangers. Do not leave luggage unlocked or unattended even for a moment. Learn where emergency exits are located, no matter where you are (grocery stores, malls, doctor's office, subways, any congested public area), and learn where staircases are located. Notice your immediate surroundings, be aware of heavy or breakable objects that could move, fall, or break in an explosion, especially store windows and shelving.

Preparing for a building explosion:

The use of explosives by terrorists can result in collapsed buildings, fires and secondary explosions due to gas, oil or other combustibles being in the building. People who live or work in a multi-level building should review emergency evacuation procedures and learn the fire routes. Know where all the exits are located. Speak with the building superintendent about keeping fire extinguishers in working order, know where they are located and how to use them.

Learn first aid. Contact the local chapter of the Red Cross for additional information.

The following emergency items should be kept in a designated place on each floor of the building:
- Portable battery operated radio and extra batteries
- Several flashlights and extra batteries
- First aid kit and manual
- Several hard hats
- Fluorescent tape to rope off dangerous areas
- Masks and whistles.

Bomb threats:

If you receive a bomb threat, get as much information from the caller as possible. Keep the caller on the line and record/write down everything that is said. Notify the police first, and then building management. Evacuate the area completely.

If there is no warning and you see a suspicious package, do not touch it. Clear the area and notify the police immediately that you have found something suspicious. When evacuating a building, avoid standing in front of it, near windows or other potentially hazardous areas. Do not stand around on the sidewalk or street. Go at least one block up wind from the area.

After an explosion:

After a building explosion, get out of the building as quickly and calmly as possible. Stay alert because the building could still be moving and walls, ceilings, wires and other debris could fall on you. If there is a fire, stay low to the floor and cover your nose and mouth with a cloth, preferably a wet cloth. Heavy smoke and poisonous gases rise first towards the ceiling. Stay below the smoke at all times.

When approaching a closed door, use the back of your hand and forearm to feel the lower, middle, and upper parts of the door. If it is hot to the touch, do not open the door. If it is not hot, brace yourself against the door and open it slowly, ready to slam it shut again if you see smoke or fire. Seek an alternative escape route.

Do not attempt to rescue people who are inside a collapsed building. You may get trapped yourself. If you are trapped, use a flashlight, not a lighter to see. Cover your mouth with a handkerchief or clothing. Tap on a pipe or wall so that rescuers can hear where you are. Shout only as a last resort. You may breathe in dangerous amounts of dust.

Chemical agents:
Chemical agents are poisonous gases, liquids, or solids that have toxic effects on people, animals and plants. Severity of injury depends on the type and amount of the chemical, and how long you are exposed to it.

If a chemical attack were to occur, authorities would most probably instruct you to shelter where you are and seal the premises immediately. An above ground location is preferable because some chemicals are heavier than air. They may seep into basements, even if the windows are closed. Using duct tape, seal all the cracks around the door and any vents into the room where you shelter.

Authorities would rarely advise you to evacuate. Leaving the shelter to rescue or assist victims can be a deadly mistake. There is no assistance that you can offer that would help if you are not trained.

PLEASE READ THE PREVIOUS SECTION ABOUT MAJOR CHEMICAL OR BIOLOGICAL EMERGENCIES ON PAGE 69

Biological agents:
Biological agents are organisms or toxins that have illness producing effects on people, animals and plants. Because they cannot usually be detected and may take time to grow and cause a disease, it is almost impossible to know that an attack has occurred. If officials become

aware of an attack through an informant or a warning by terrorists, they would most likely advise everyone to stay where they are and seal the premises.

Those affected by a biological agent require immediate attention of specially-trained personnel. Some agents are contagious and an affected person needs to be quarantined. Whether you are directly affected or not, wear a mask and protective clothing.

PLEASE READ THE PREVIOUS SECTION ABOUT MAJOR
CHEMICAL OR BIOLOGICAL EMERGENCIES ON PAGE 69

Biological/radiological exposure:

Listen to local radio and television reports for the most accurate information from the government and medical authorities on what's happening and what actions you need to take. If you have been directly affected, you will have to shower with soap and water immediately, then get to a medical facility, unless the authorities want you to wait until the cloud of toxins has passed.

Wear protective clothing and a mask. Cover as much skin area as possible. If you suspect you have been biologically infected and later find out you have not, the mask and clothing will give you some protection.

Depending on whether you are contagious, there will be different procedures to follow.

If you suspect you are contagious, cover your body and face as best you can so that you will not infect anyone on your way to, or inside the hospital. You may have to call the police or an ambulance to come and get you from wherever you are so that you don't infect anyone outside the infected zone. Make only one call to your family contact, to advise him/her quickly of what is going on, then keep the phone lines free.

IMPORTANT: PLEASE READ FIRST AID
SECTION FOR EXPOSURES ON PAGE 83

IMPORTANT: PLEASE READ GENERAL
SAFETY PROCEDURES ON PAGE 89

REVIEW

- Be alert and aware of your surroundings.
- Learn to automatically note where emergency exits are located, wherever you are.
- Review all emergency procedures where you work.
- Have a portable radio and spare batteries on hand.
- If you receive a bomb threat, write down everything that is said. Call 911, then management if you are in a building.
- Evacuate as fast as possible.
- If there is a bomb explosion, check closed doors for fire on the other side before opening them.
- Do not attempt to rescue trapped people.
- If you are trapped, cover your nose and mouth. Tap on pipes or walls. Shout only as a last resort.

If there is a chemical/biological attack:
- Seal the area you are in, unless told to evacuate or move to a higher level in the building. Seal that new area.
- Do not leave your area to assist others.
- Wear protective clothing and a mask.
- Monitor the radio/TV constantly for instructions.
- Call for medical assistance immediately since the injured may need to be decontaminated.
- Read the chapter outlining steps to take for basic first aid.
- Call for medical assistance and advise the medics that you may be infected so they can protect themselves en route to the hospital and arrange quarantine for when you arrive there.

21 | If Disaster Strikes

Try to stay calm as your attitude will affect others, especially children.

The danger is not necessarily over once the emergency has passed. You and your family may have to evacuate an area. If you are asked to leave, do so immediately.

Wear long-sleeved shirts, long pants and sturdy shoes.

Take your emergency supplies kits.

Take your pets and their kits.

Call your family contact to tell him/her where you are going and when you expect to arrive.

Do not use the phone after that as the lines must be kept open for emergency personnel use.

Turn off water and electricity before leaving, if instructed to do so. Leave natural gas service ON unless local officials advise you otherwise. You may need gas for heating and cooking later, and only a professional can turn it back on. It could take weeks for a professional to respond.

Lock your home.

Do not visit the disaster area, as you may hinder rescue or recovery efforts.

Do not return home until local authorities have advised that it is safe to do so.

Monitor local media reports for advice and the safest route home. Do not take short cuts.

Drive carefully and watch for debris, dangling or broken wires, damaged bridges and roads.

Report any problems to your local police or fire department.

Wait for the all-clear signal before you enter buildings that have been structurally damaged.

Be aware that heavy law enforcement by police and military at local, provincial/state and federal levels follow a terrorist attack. Health and mental health resources in the affected communities can be stretched to their limits. Extensive media coverage, public fear, and other consequences can continue for a long time. Workplaces and schools may be closed and there may be restrictions on travel. Remember that clean-up may take months.

If you are told to shelter in place, stay home and do the following:

Close and lock all windows and exterior doors. Close the fireplace damper.

Turn off all fans, heating and air conditioning systems.

Get your emergency supplies kit.

Turn on the radio or TV.

Go to an interior above ground room without windows.

People who may have come into contact with a biological or chemical agent may need to go through a decontamination procedure and receive medical attention.

If you encounter someone who is injured, check to make sure it is safe for you to approach. Then check the victim for unconsciousness and life threatening conditions. Someone who has life threatening conditions, such as not breathing or severe bleeding, requires immediate care by trained personnel. Call out for help.

IMPORTANT: PLEASE READ GENERAL
SAFETY PROCEDURES ON PAGE 89

REVIEW

- Do not visit the disaster area.
- Do not return home until advised by authorities that it safe to do so.

22 | First Aid

THOROUGHLY WASH YOUR HANDS WITH SOAP AND WATER
BEFORE AND IMMEDIATELY AFTER GIVING CARE.

Reduce any care risks. Avoid direct contact with blood and any
other body fluids. Use protective equipment such as disposable gloves
and breathing barriers such as masks, rags, or handkerchiefs.

There are steps you can take to care for someone who is hurt but
whose injuries are not life threatening.

Control bleeding:

Cover the wound with a dressing and press firmly against the wound
if there is nothing inside it. If there is something inside, it is NOT
always good to try to remove it as you may end up pushing it deeper
and causing more damage. If you remove a large or deep object, you
may cause more bleeding. The best course may be to call for help.
Elevate the injured area above the level of the heart if you do not suspect that the person has a broken bone. If the bleeding does not stop,
apply more dressings and bandages. Watch for shock.

Shock:

The signs are weakness, confusion, pale or blue lips and nails, cold
sweat, weak and rapid pulse and shallow rapid breathing. Watch especially for the breathing. Keep the person from getting too chilled or

overheated. Loosen tight clothing. If legs are not broken, elevate them about 12 inches (30 cm). Do not give food or water, only wet victims' lips. If they vomit, turn them sideways so they don't breathe in the vomit and choke. Get medical attention fast.

Burns:

Wash your hands and, if possible, put on clean gloves. Stop the burning. You can do this by pouring large amounts of water over the area. DO NOT immerse it. Do not use ice. Watch for shock. Elevate the burn area. Cover the burn with dry, clean dressings. Do not use plastic. Do not apply creams or ointments on the burn. Do your best not to touch it.

Broken bones, joints or muscle injuries:

Watch for shock. Cover the person to avoid chills. Clean your hands and, if possible, put on clean gloves. Cut away the clothing so you do not move the broken area. If the area is bleeding, try to stop it without pushing on the bone. Cover it with clean cloth or gauze. Keep pressure on the bleeding area if possible. Make a splint with anything handy. Tie it carefully and gently above and below the break if at all possible. Make sure it is not too tight. Tie up broken arms at a 90-degree angle upwards against the chest. Try to splint broken legs and tie the legs together to keep them still. Never try to set the bone yourself by pushing the ends back together; that needs to be done by trained professionals. Apply ice or a cold pack for a few minutes at a time to control swelling and reduce pain. Avoid any movement or activity that causes pain unless the scene becomes unsafe and the injured person needs to be moved. Call for immediate help.

Prickly heat rash:

It comes from blocking of the sweat ducts. Trapped sweat cannot reach the surface and escape. The sweat ducts look like pinhead-sized holes that become itchy, prickly, and have a burning feeling. Sometimes you can get a secondary infection and tiny pustules may form. Take a shower or cool bath. You can use a light dusting powder. Do not use creams or lotions, as they can block the sweat ducts. Wear light, loose clothing. The rash will disappear in a cool environment after a while. How long that takes depends on the person, how bad the rash is and environmental conditions. If you get an infection, you should see the doctor for advice.

Heat cramps:

Symptoms are muscle spasms, extreme thirst, nausea, cold clammy skin, and often heavy sweating. Move the victim to a cool, shaded area to rest. Apply firm pressure, warm wet towels or hot water bottle to cramping muscles. Give the victim 2 glasses of lightly salted water at 10 to 15 minute intervals if the cramps persist. (Mix 1 teaspoon/ 5 ml of salt into 1 quart/1 liter of water).

Heat exhaustion:

Symptoms could be any or all of the following: sweating, weakness, headache, dizziness, fainting, scant urine, low blood pressure, disorientation, vomiting, muscle cramps or spasms, cold and clammy skin, and body temperature that is slightly lowered or elevated. Move the victim to a cool area to rest. Loosen or remove clothing. Provide cool sips of salty water – no ice – (1 teaspoon/5 ml of salt into 1 quart/ 1 liter of water) or sports drinks containing electrolytes, and cover the person if shivering. The victim should rest in bed until recovered. Give stimulants such as coffee or tea. Stop all fluids if the victim begins to vomit. If the person faints, has confusion or seizures or if body temperature rises over 104 degrees Fahrenheit (40 degrees Celsius), seek medical help immediately.

Heat stroke or sun stroke:

This is a grave emergency that can be fatal. Symptoms are weakness, headache, hot and dry skin, no sweating, dilated pupils, offensive body odor, nausea, rapid breathing, and sharply rising temperature. A pounding pulse and full, elevated blood pressure, delirium, or coma, are all common occurrences. Skin may be flushed at first, later pale white or purplish. Call for an ambulance immediately.

While you are awaiting medical help, you must act quickly to cool the person. Move to a cool shaded area. Apply an ice bag, such as crushed ice in a cloth, or cold cloths to the head. If the person cannot be moved to a shelter, drench his or her clothes with cold water, poured on or sprayed from a hose. Preferably strip off clothing and wrap the victim in a cold wet sheet. Have electric fans blow on the cold sheet. Keep the cold cloths or ice on the head. Check the body temperature every 10 minutes or so. Continue treatment until the temperature drops to 101–102 degrees Fahrenheit (38–39 degrees Celsius).

Signs of dehydration:

Dry or sticky mouth, dry eyes with no tears, lack of urination, dry cool skin, irritability and listlessness, tiredness or dizziness. You can get dehydrated by losing body water (which is sweat) from high fever, diarrhoea, vomiting, sweating a lot on a hot day or from excessive exercise. It can usually be treated by simply drinking fluids. If faintness keeps on happening when you stand up, even after several hours, or you have very little urine output, you need to see a doctor.

Signs of frostbite:

- First degree: burning or throbbing pain, partial skin redness, usually no blisters.
- Second degree: redness and blisters appear. The blisters can become blackened, with numbness.
- Third degree: victim has no feelings in the affected area, no burning, throbbing or pain.
- Fourth degree, the most severe: all the skin, muscle, tendons and bones are affected. There is very little swelling, but there is pain at the joints.

Frostbite can permanently damage the skin.

Watch for signs of hypothermia, which can accompany frostbite. See below for treatment for hypothermia.

Hypothermia:

This happens when your body temperature drops so much that the body can no longer generate as much heat as is being lost. Confusion, slurred speech, stiff muscles, or uncontrollable shivering are symptoms. If they occur, seek medical assistance immediately. Hypothermia can be fatal.

If frostbite or hypothermia is suspected, begin warming the person slowly and seek immediate medical assistance. Warm the person's trunk first. Use your own body heat to help. Arms and legs should be warmed last because stimulation of the limbs can drive cold blood toward the heart and lead to heart failure. Put the person in dry clothing and wrap his/her entire body in a blanket. Never give a frostbite or hypothermia victim something with caffeine in it (such as coffee or tea) or alcohol. Caffeine, a stimulant, can cause the heart to beat faster and hasten the effects extreme cold has on the body. Alcohol, a depressant, can slow the heart and also hasten the ill effects of a cold body temperature.

For frostbite, wrap the frostbitten area in dry, sterile or clean cloth. Elevate the area and warm it. If you are outdoors and must re-warm an area, you can use water warmed over a fire to about bathtub temperature (about 104 degrees Fahrenheit or 38 degrees Celsius). The person must then be brought immediately to professional care. In the meantime, do not let the injured part re-freeze.

If you are days away from emergency assistance, do not use the water to re-warm the area if there is a possibility of it refreezing. Separate fingers and toes with cotton or material before complete wrapping them. Get help as soon as possible.

Chemical poisoning:

The symptoms include difficulty breathing, changes in skin color, headache or blurred vision, dizziness, irritated eyes/skin/throat, unusual behavior, clumsiness or lack of coordination, stomach cramps or diarrhoea. If a child eats or drinks a non-food substance, find the container and bring both of them to the phone. Call the poison control center or 911. Follow their instructions carefully. Do not give anything by mouth until you have been advised to do so by medical professionals.

If you think you have been exposed to a toxic chemical, call the poison control center or 911. If you see or smell something that you think may be dangerous or find someone who has been overcome with toxic vapors, your first job is to make sure that you don't become a victim. Call or send someone else to call 911 right away.

Chemical and biological emergencies:

A chemical burn can be minor or life threatening, but proper treatment can reduce the chance of infection and the damage. Unfortunately, there are some chemicals that are not compatible with water. In those cases, the effects can be worse.

It is a chance that you take using water, but without knowing the type of chemical or having official medical advise, you don't have much of a choice unless you are in or near a chemical plant that has specific emergency supplies handy.

Remove any affected clothing or jewelry from the injury. Use lots of cool running water to flush the chemical from the skin until emergency help arrives. The running water will dilute the chemical fast enough to prevent the injury from getting worse. Take care not to contaminate other areas of skin or other people with the clothing that has been removed. Place the clothing in a plastic bag so it cannot contaminate anything else.

Use the same treatment for eye burns. Remove contact lenses. Be careful to flush the eye from the nose outward.

If there isn't any clean water available, gently brush the chemical off the skin and away from the victim and you. If the chemical is on the face, neck or shoulders, ask him victims to close their eyes before brushing off the chemical. Cover the wound loosely with a dry, sterile or clean cloth. Do not put any medication on the wound. Seek medical attention immediately.

Biological/radiological contamination:

If you have been directly affected, you will have to shower with soap and water immediately, then get to a medical facility, unless the authorities want you to wait until the cloud of toxins has passed. Depending on whether you are contagious, there will be different procedures to follow. Wear protective clothing and a mask. Cover as much skin area as possible. If you suspect you have been biologically infected and later find out you have not, the mask and clothing will give you some protection. If you suspect you are contagious, cover yourself the best you can so that you will not infect anyone on your way to, or inside the hospital.

23 | General Safety Procedures Following Disasters

Listen to your radio for emergency instructions. Try to remain calm and patient. Your attitude will affect others around you, especially children. Panic leads to mistakes.

Wear protective shoes and clothing, as there will be a lot of glass and debris that may still come down and be lying around. You also need to wear rubber gloves and boots and protective eyewear when cleaning up.

If you are in the clear and there is no danger to you, help locate and treat anyone who might be injured. Help neighbors who require special assistance with infants, the disabled, large families and elderly people. Confine and secure your pets.

Call your family contact. Do not use the telephone again unless there is a life-threatening emergency.

You may have to evacuate a building. It is better to get out and stay out even if everything looks OK. Do not return until the authorities give permission, especially for buildings that have been structurally damaged.

Electricity:
If you are inside and it is safe, look for electrical damage.

USE A FLASHLIGHT. Do not strike a match or use an open flame unless you know the gas is turned off. Check for blown fuses and look for short circuits in the wiring and equipment. If you see sparks, broken or frayed wires, or if you smell burning insulation and you suspect the electrical lines have been disrupted, and if it's possible to do so, turn off your electrical power at the main fuse box or circuit breaker by using a piece of wood.

If you are NOT wet or standing in water and the electrical appliances are wet, turn off the main power switch. If you have to step into water to get to the fuse box or circuit breaker, call an electrician first for advice.

Unplug the electrical appliances and let them dry before checking for damage. Electrical equipment should be verified by a professional before being returned to service. When power does return, do not turn on everything at once. Reconnect the main power supply and connect large appliances one every hour. Do not plug in your computer for at least 24 hours as there will be power surges that can fatally damage it. If the fuses blow when the power is restored, turn off the main power switch again as directed above and call the electrician. Do not operate electric appliances, like vacuum cleaners, while standing on wet carpet or floors, especially not on wet concrete.

Contact the furnace maintenance company to inspect the furnace and chimney.

Look for fire hazards.

Gas lines:

There may be broken or leaking gas lines. If you smell gas or hear a blowing or hissing noise, open a window, turn off the gas if you can, and quickly leave the building. Call the gas company from a neighbor's home. If you turn off the gas for any reason, a professional must turn it back on.

Water damage:

Before entering a flooded building, check for foundation damage and make sure all porch roofs and overhangs are supported. Check for sewage and water line damage. If you suspect sewage lines are damaged, avoid using the toilets and call a plumber. If water pipes are damaged, contact the water company or municipal services and avoid using water from the tap. You can obtain safe water from undamaged

90

water heaters or by melting ice cubes. Do not use your regular water supply or septic system until it has been inspected and declared safe for use. Wells that have been flooded should be tested for bacteria and should be disinfected after floodwaters recede. Use tap water only if local health officials advise it is safe.

Dispose of all contaminated food. You can clean cans and sealed containers with bleach if they are intact, rinsing very well before opening them.

Mold:

Mold can cause serious health problem, including allergies and asthma. If you discover mold in your home, you must destroy it quickly. Bleach is the only product that kills it on contact. However, be very careful handling this product. Use 1 part bleach to 10 parts water to disinfect. Inspect your food before preparing it. Dispose of all contaminated food. You can clean cans and sealed containers with bleach if they are intact, rinsing very well before opening them. Freeze valuable books and documents to retard mildew growth until proper drying can be performed.

24 | **Emotional Reactions**

Pay attention to your feelings and those of your family members. You won't feel or act like yourself for a while. Depending on the severity of the disaster, reactions can range from feeling faint, confused, and trembling, to numbness and vomiting. After the event is over, people often feel bewildered, shocked, angry, threatened, scared, insecure, and relieved to be alive. Later, many people sleep poorly, have nightmares, lose their appetite, or panic at the slightest hint of a storm or other emergency.

Children may start thumb sucking, bedwetting, or acting very young and using baby talk after the disaster. They may get stomach aches. They may be afraid the disaster will happen again, or that they or you might get hurt. They may also fear that they will become separated and lose you. All of these reactions are considered normal.

Children frequently misinterpret what is going on, or what has actually happened. Often events become magnified and distorted in their minds. They need to express their feelings, either by talking, drawing, or play-acting. They must be comforted and reassured. Give children honest, plain, and simple explanations. Encourage children to talk and ask questions about the disaster and the people affected.

You also need to talk with your peers about your feelings. Give you and your family time to grieve and heal. Local health services are usu-

ally able to provide assistance in coping with the mental trauma result-
ing from a disaster.

25 | Bi-annual Checklists

As time goes by, children grow and medical conditions change, food freshness and medicine expire. There may also be changes with family members or pets. Go through your kits and lists in the spring and the fall to replace clothing that is too small, too big, or not appropriate for the season's weather. You may need to change the types or quantity of food and water required, especially with growing children. As children grow, their toys, games, music, books, crayons, and coloring books become no longer appropriate. These need to be changed too.

Some phone numbers may need to be added, changed, or deleted on your master list. In some areas, roads and bridges may have been added or closed. Verify escape routes. Some cities may also change the location of local evacuation shelters.

It is not a big task to go through your kits and lists since everything is already in place. It is just a matter of phoning, verifying, and exchanging some things. Check off the items or write down the date when you verify the items. That way you will save time and know at a glance what is done or what needs to be done and when.

Check the following items and write down the date they were verified.
  – Clothing and footwear sizes
  – Medicine kit expiry dates or adjustments
  – First aid kit expiry dates

- Food freshness expiry dates
- Food requirements or adjustments
- Water supply expiry dates and quantity
- Children's activities
- Phone numbers
- Home safety supplies
- Car safety supplies
- Office safety supplies
- Escape routes
- Shelter locations.

Life is precious and fragile.

I cannot urge you strongly enough to take charge of your own safety and survival.

I hope that all you have read here will help you prepare for any emergency and if one comes your way, you will get through it safely.